My Dog Ate My Calculator

Using the TI-84 for Introductory Statistics

Isabella Romeo

Sherwood Forest

http://CalculusCastle.com

My Dog Ate My Calculator: Using the TI-84 for Introductory Statistics, First Edition, by Isabella Romeo
Last Revised on: 1/4/18 (Version 1.0)
Sherwood Forest Books, Los Angeles, CA, USA
ISBN-13: 978-0996686037 (Sherwood Forest Books)
ISBN-10: 0996686037

Statistics books suck. Don't expect this one to be any better. Please report any errors, omissions, or suggestions for improvements through the form at `https://github.com/bellaromeo/faqsinstats`.

Table of Contents

Preface

This small book is intended to help students remember (or learn for the first time) the keystrokes needed to complete simple statistical calculations with the TI-84.

You will not learn statistics by reading this book. This book can be used to accompany any introductury book in statistics, even if that book does not refer students to the use of a statistical calculator.

You will not learn how to use the TI-84 in general here. The TI-84 is a general purpose scientific and graphing calculator, and does far more that just solve statistics problems.

You will learn how to solve, with the calculator, the typical problems encountered in a freshman statistics class. This book is best used as a lab reference or manual to accompany introductory textbook in statistics.

Readers are asked to report any errors, omissions, or suggestions for improvements at `https://github.com/bellaromeo/faqsinstats/issues/new` .

1. Putting Data into a List

To enter a sequence of numbers into a list:

1. Select stat EDIT, then scroll down to **`1:Edit...`**, and press enter.

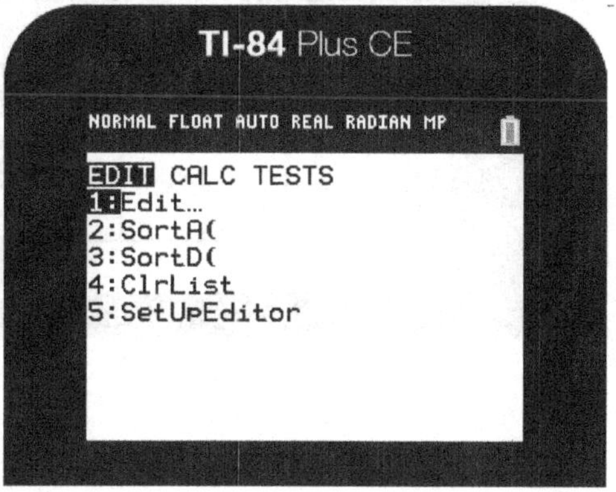

2. Using the arrow keys, scroll to the top of the list you want to use.

3. If the list you want to use already contains data you can clear the list by moving the cursor to the label at the top of list (e.g., **`L1`**, **`L2`**, etc.), and then pressing (in sequence) the keys clear enter.

4. Enter the data items into the list **`L1`** (or your desired list).

   Use the arrow keys to navigate through the lists.

Type [enter] after each item.

If you make a mistake, you can use the arrow keys, and/or the [del] (delete) or [ins] (for insert) keys.

.

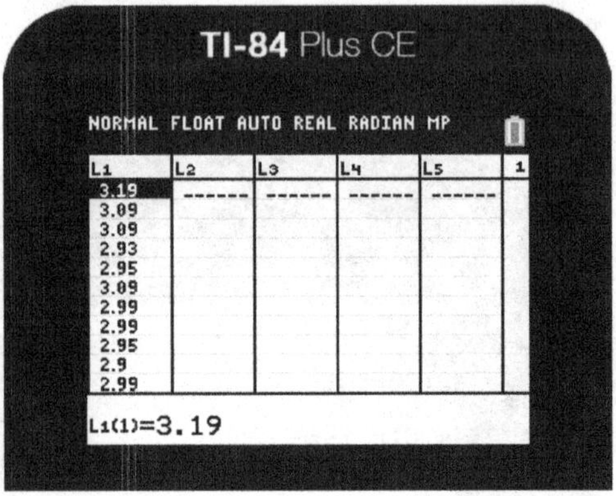

2. Mean, Standard Deviation, Median, and Quartiles

1. Put the data into a list (see chapter 1).
2. Press stat then CALC. Scroll down to **1:1-var-stats** and hit enter.

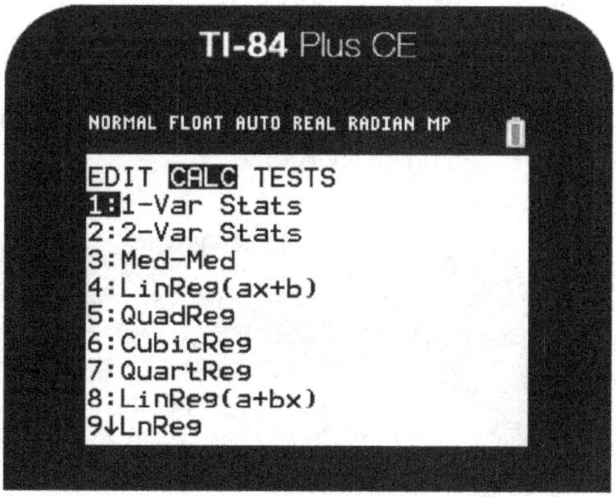

On the next screen enter:

List: **[L1]**
FreqList: *(leave blank)*

Scroll down to **Calculate** and press enter.

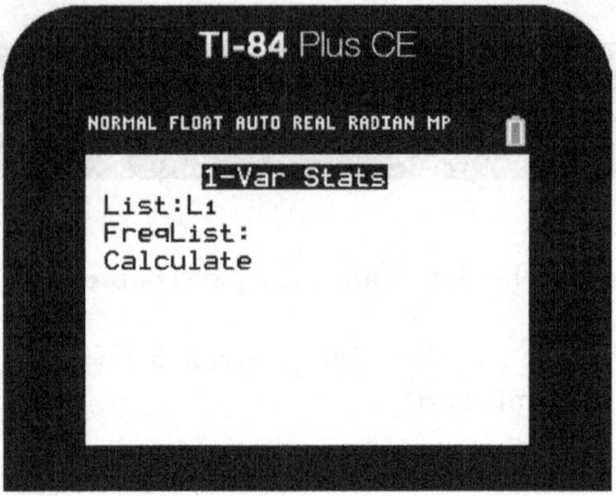

3. The answer screen gives a lot of information. The mean is given on the line labeled $\bar{x}$. The standard deviation is on the line labeled **Sx**. Quartiles, median, maximum, and minimum are also shown (you have to scroll off-screen).

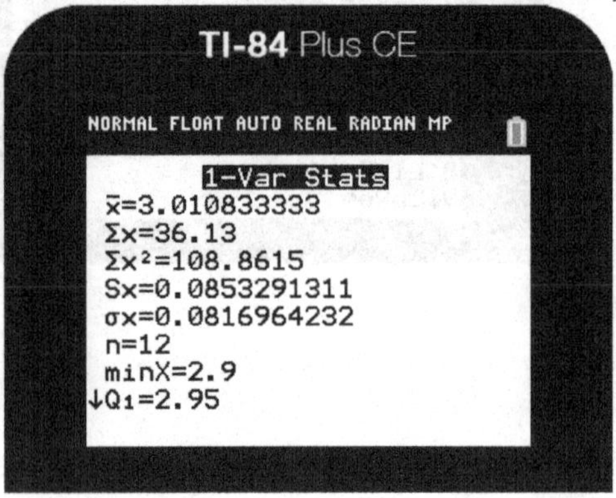

3. NormalCDF - Area Under a Normal Distribution

> ***Example.*** The SAT has a mean of 500 and standard deviation of 100. Find the proportion of scores between 450 and 600.[1]

1. Sketch the area that needs to be found, and determine the **upper** and lower bounds of the plot. It will look like one of the following three plots.

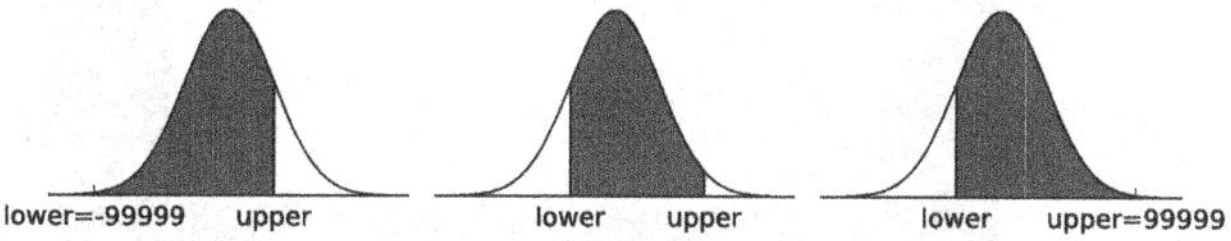

In the SAT example, the plot looks like the one in the middle (with **lower=450** and **upper=600**). In general the result will fall into one of three cases:

Problem Given	Use for **lower**	Use for **upper**
Between *lower value* and *upper value*	*lower value*	*upper value*
Below *upper value*	**-999999**	*upper value*
Above *lower value*	*lower value*	**999999**

Here the sequence **999999** can be replaced by **1E99** and the value **-999999** by **-1E99**. The point is to

[1] This is identical to the following problem. The SAT has a mean of 500 and standard deviation of 100. Find the probability that a single student chosen at random scored between 450 and 600.

use a number that is much larger in magnitude than any other number in the problem, and the string of **9**'s is much easier to type in.

2. Select 2nd distr (distr is above the vars key), scroll down to **2:normalcdf**, and hit enter.

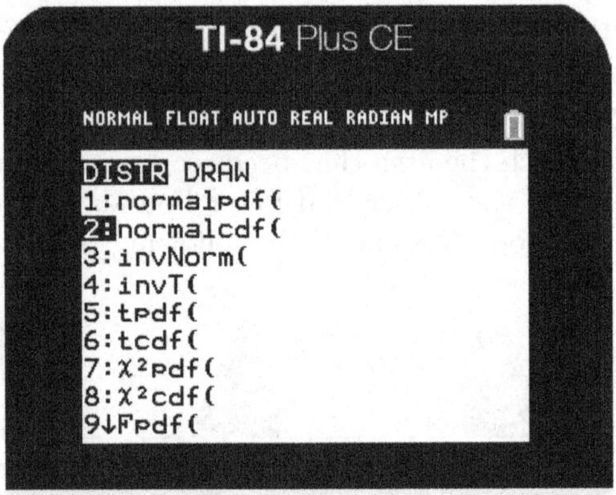

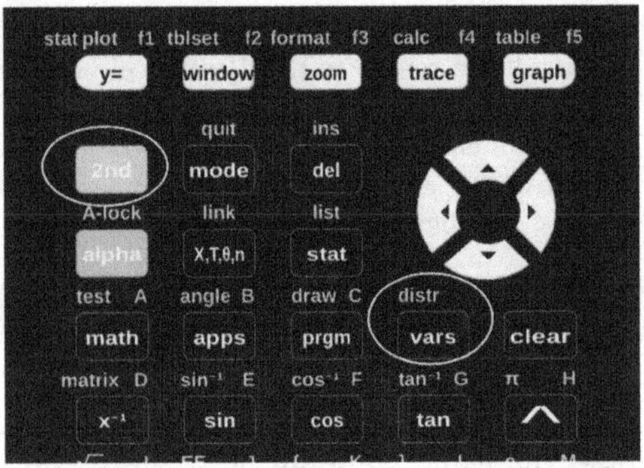

3. On some calculators, a menu will appear. If a menu does not appear, skip to 5.

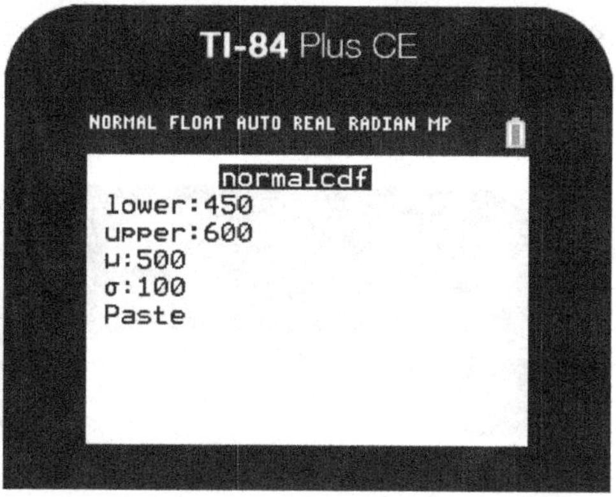

The **lower** and **upper** values are the left and right x-value (endpoints for the area under the curve). μ and σ are the mean and standard deviation. You should fill in the correct values for these numbers. Then scroll to **Paste** and click on enter. The screen will display something like this:

normalcdf(450, 600, 500, 100)

Skip to step 6.

4. If a menu does not appear (as in the previous step), you should see the following on your screen:

   normalcdf(

   The open parenthesis tells you that you have to fill in the rest of the information. The syntax is

   normalcdf(lower, upper, μ**,** σ**)**

   where **lower**, **upper**, μ, and σ are as above.

   To solve the SAT problem, you will need to add the values of all the input. In this case we must enter 450, 600, 500, 100 and the final closing parenthesis, all separated by commas, and then press enter.

normalcdf(450, 600, 500, 100) enter

5. After pressing enter, the answer is displayed.

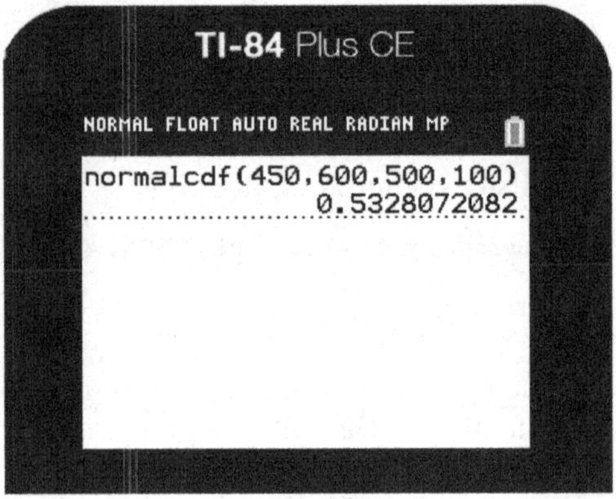

In the example, the area is around 0.5328. Using the original language of the problem, the proportion of students who score between 450 and 600 is around 0.5328 (or 53.28%).

4. invNorm – Inverse Normal Distribution

Example. The SAT verbal exam has a mean of 500 and standard deviation of 100. A certain school will only accept students who score at least at the 90th percentile. What is the minimum verbal SAT score x required for entrance to that school?

The 90th percentile means the value of x such that that 90% of students score lower than x. If we plot the distribution function $N(500, 100)$, the area to the left of x is 0.9. Equivalently, the proportion of students scoring less than or equal to x is 0.9, and the probability of a single student chosen at random of scoring below x is 0.9.

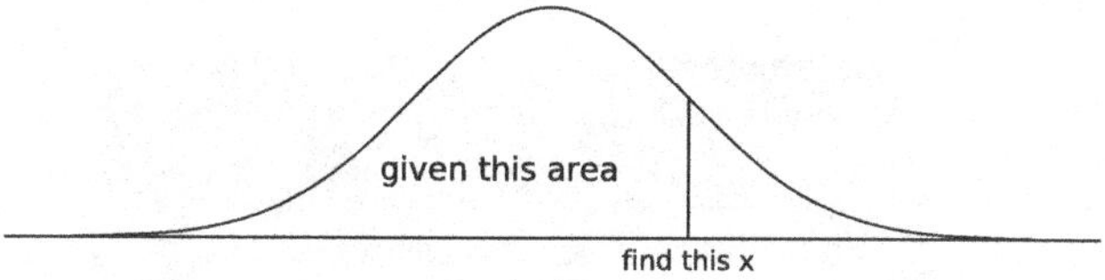

Our goal is to find x given that the area to the left of x is 0.9, using the **invNorm** function of the TI-84.

1. Select `2nd` `distr` (`distr` is above the `vars` key).

2. Scroll down to **3:invNorm**, and hit `enter`.

3. On some calculators, a menu will appear. If a menu

does not appear, skip to 5.

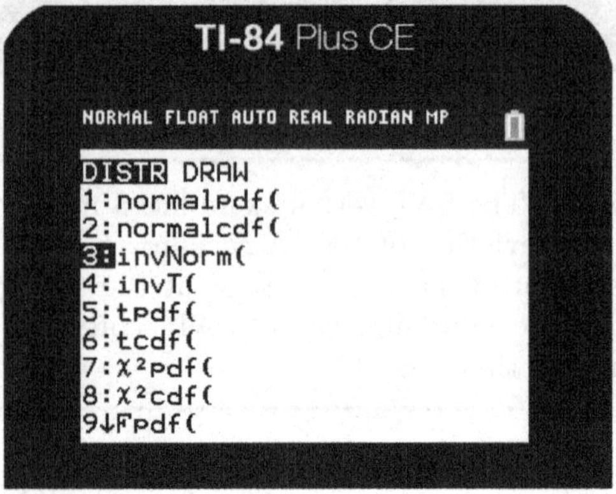

4. If the menu appears, fill in the values for **area** (in this case, 0.90), μ (the mean, in this case it is 500), and σ (the standard deviation, in this case it is 100). Then scroll to **Paste** and hit enter twice (the first to paste, the second to get the answer). Continue with step 6

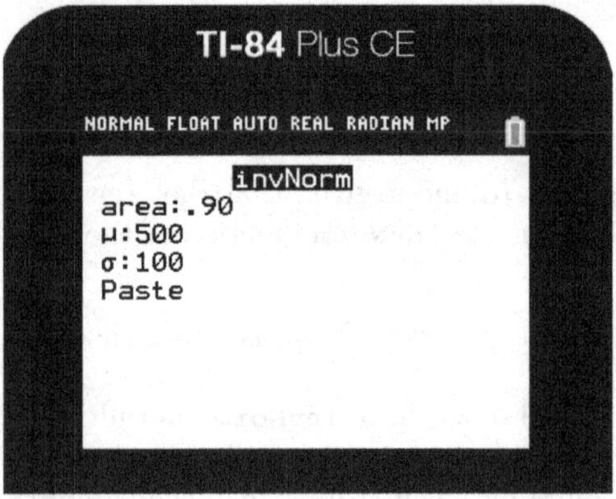

5. If there was no menu, then the following will have been pasted onto your screen:

   invNorm(

   You need to fill in the rest. The general syntax is

   invNorm(area, μ, σ)

   where **area**, μ and σ are as defined above. Follow this with the [enter] key twice. For this example,

   invNorm(0.90, 500, 100) [enter]

6. After the second [enter], the answer will be displayed on the screen.

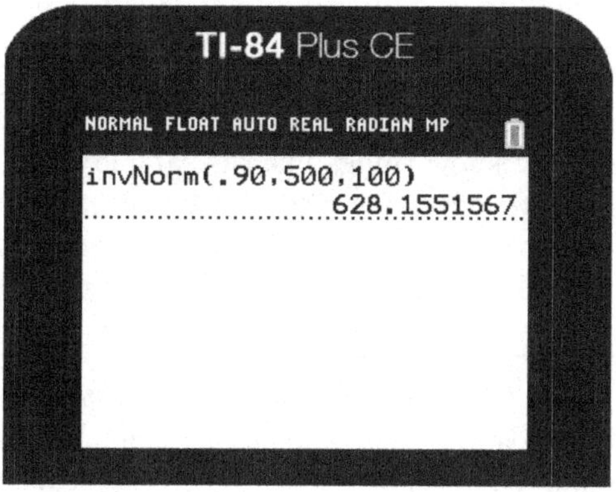

In the example, the 90th percentile falls at around 628. In terms of the original language of the example, only students with a score above 628 will be admitted to this particular school.

5. T-Test – Hypothesis Test (from Data)

You can compute many of the tests in the TI-84 either from data or from statistics.

When you compute a t-test using data, that means you are starting from a list of data values. The procedure for this is described in the present chapter. If you already know the sample mean and standard deviation, you should use the procedure in chapter 6 instead of the procedure in this chapter.

When you compute a t-test using statistics, that means you already know things like the sample mean and sample standard deviation. You do not need to have a list of the individual data values. The procedure for this is described in chapter 6.

> ***Example.*** A professor designs an exam for students that should take about an hour but she is not sure that it can be completed in time. She picks 5 typical students to take the exam, and times how long it takes them to finish it. Their times (in minutes) are: 37, 42, 53, 67, and 78 minutes. She wants to test the hypothesis $H_A : t > 60$ against the null hypothesis $H_0 : t = 60$ to see if the test is too hard.

1. Enter the data values into list `L1`. Do this by typing `stat` `edit` and then typing the data values into the column with the `L1` at the top of it. (Another

list may be used if you prefer.) The screen will look something like this when you are done:

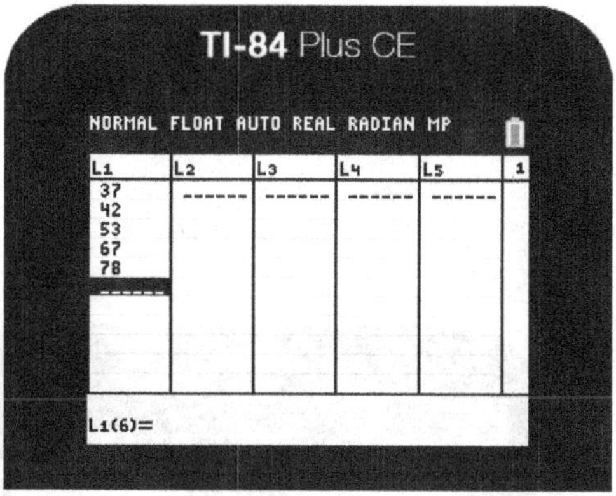

2. Select [stat] [TESTS] and scroll down to [2:TTest]. (you may have to scroll off screen to the next page) Press [enter].

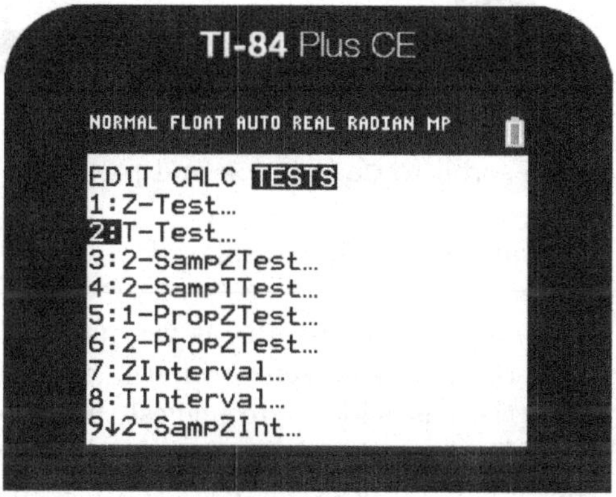

3. On the first line, that says

Inpt: Data Stats

scroll to **Data** and press [enter] before proceeding.

Enter the following:

μ_0 : **0**
List: L1
Freq: 1
$\mu :\neq \mu_0 < \mu_0$ $\boxed{> \mu_0}$

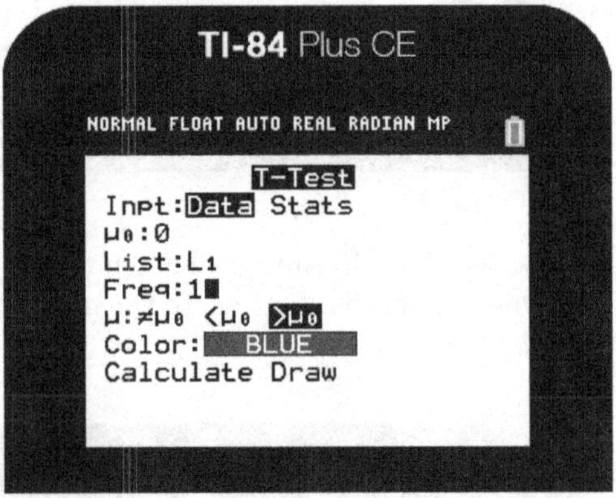

Then scroll to **Calculate** and press [enter].

4. The output screen is displayed.

   For this data, the p-value is very small. As we see from the following screen image, $p = 0.00096297$. Thus there is evidence to suggest that the professor should accept the alternate hypothesis and reject the null hypothesis at, e.g., the $p = 0.001$ level. In the language of the original problem, the test is too hard for most students to complete in an hour.

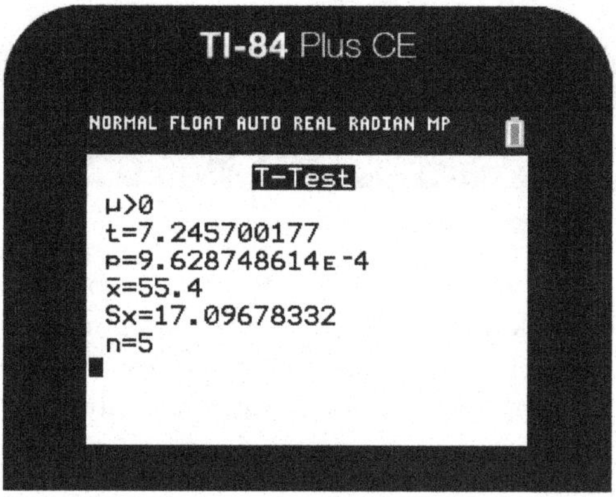
TI-84 Plus CE
NORMAL FLOAT AUTO REAL RADIAN MP
T-Test
μ>0
t=7.245700177
p=9.628748614E-4
x̄=55.4
Sx=17.09678332
n=5

6. T-Test – Hypothesis Test (from Statistics)

This chapter covers the procedure to perform a t-test when you know the sample mean and sample standard deviation. If you do not know these values, but you do have the original collection of data values, you should use the procedure in chapter 5 instead.

Example. A car manufacturer claims that its vehicles average 24 miles per gallon. A testing organization believes that the vehicles do not perform this well, and runs their own tests on 5 vehicles. They find an an average performance of $\overline{x} = 22.8$ mpg with a sample standard deviation of $s = 1.2$ mpg. Perform the hypothesis test of $H_A : \mu < 24$ against $H_0 : \mu = 24$.

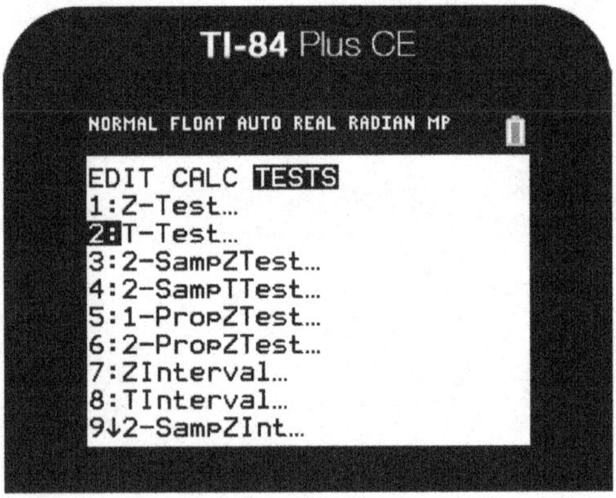

1. Select stat TESTS and scroll down to 2:TTest. (you may have to scroll off screen to the next page). Press enter.

2. On the first line, that says

   `Inpt:   Data Stats`

   scroll to **`Stats`** and press enter before proceeding. This will ensure that the statistics input menu is displayed.

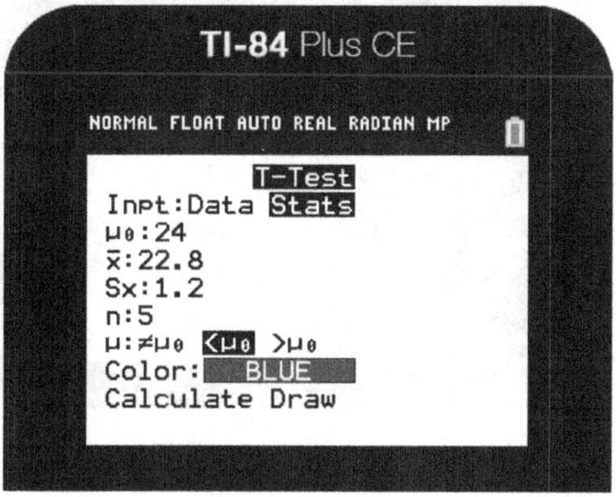

3. Enter to appropriate data on the input screen:

   μ_0 : **24**
   $\overline{x}$: **22.8**
   Sx: **1.2**
   n: **5**
   $\mu :\neq \mu_0$ $\boxed{< \mu_0}$ $> \mu_0$

   Then scroll to **`Calculate`** and press enter.

4. The results appear on the next screen.

For this example, the p-value is approximately $p = 0.0445$, as illustrated on the screen image. If we are using a significance level of 5%, then the result is significant (because $0.0445 < 0.05$). We would reject the null hypothesis and accept the alternative. In terms of the original problem wording, we would conclude that the typical gas mileage was actually less than the advertised 24 miles per gallon.

7. TInterval – Confidence Interval (from Data)

Example. A professor designs an exam for students, but is not sure how long it will take. She asks 5 typical students to take the exam, and times how long it takes them to finish it. Their times (in minutes) are: 37, 42, 53, 67, and 78 minutes. Find a 95% confidence interval on the typical amount of time it will take all comparable students to finish the exam.

1. Enter the data values into list [L1]. Do this by typing [stat] [edit] and then typing the data values into the column with the [L1] at the top of it. (Another list may be used if you prefer.) The screen will look something like this when you are done:

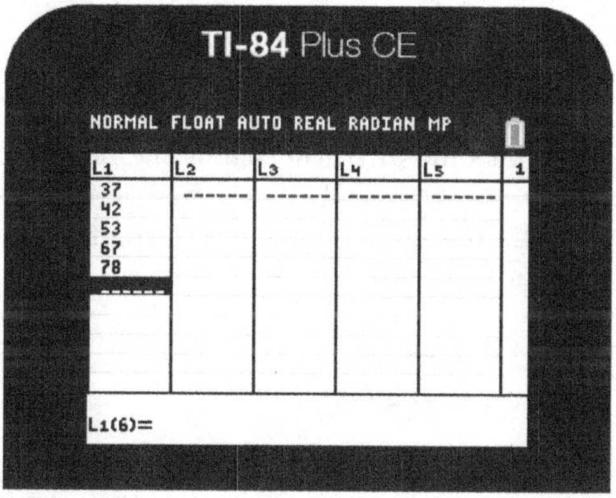

2. Select [stat] [TESTS] and scroll down to [8:TInterval]. (you may have to scroll off screen to the next page) Press [enter].

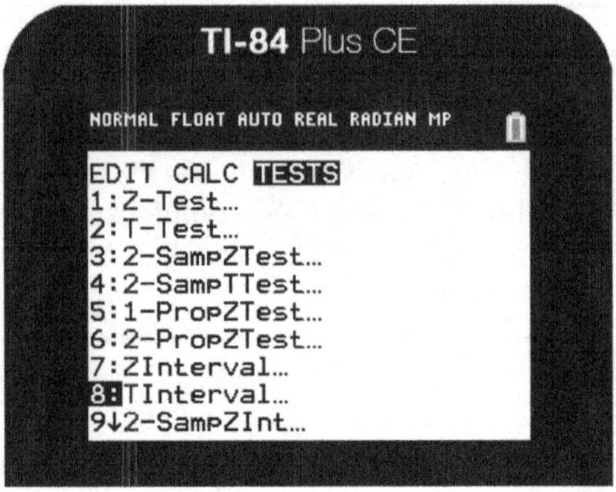

3. On the first line, that says

   `Inpt:  Data Stats`

   scroll to **`Data`** and press [enter] before proceeding. This will ensure that the rest of the menu displays the Data menu (rather than the Statistics menu).

4. On the remaining lines, enter the following:

`List:`	**`L1`**
`Freq:`	**`1`**
`C-Level:`	**`0.95`**

   Then scroll to **`Calculate`** and press [enter].

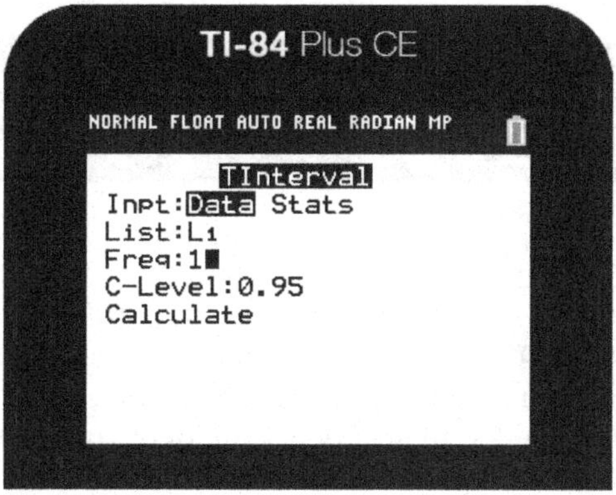

5. The output is displayed as:

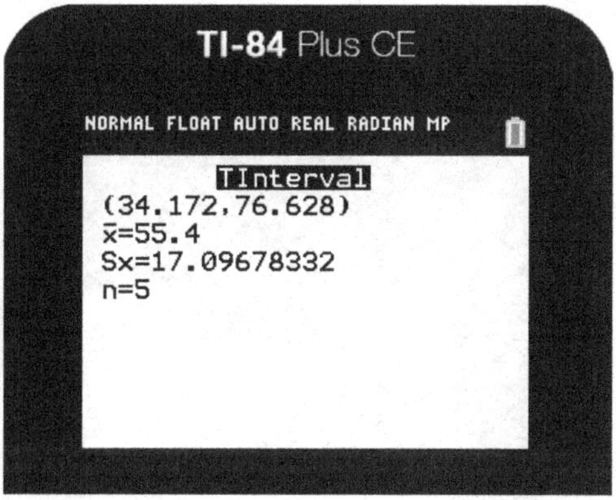

For this example, the confidence interval is [34.172, 76.628] minutes.

The center of the interval is $\overline{x} = 55.4$. The margin of error is (76.628-34.172)/2=21.228. In plus/minus notation the confidence interval is 55.4 ± 21.2 minutes (rounding to tenths of a minute).

8. TInterval – Confidence Interval (from Statistics)

Example. A professor gives an exam to a large group of students. To get an early estimate of the test results, he grades seven exams, selected at random, and determines the mean and standard deviation of of the sample scores. He finds that $\overline{x} = 63$ and $s_x = 9.2$. Use this information to estimate a 95% confidence interval on the grades for the entire class.

1. Select stats tests 8:TInterval.

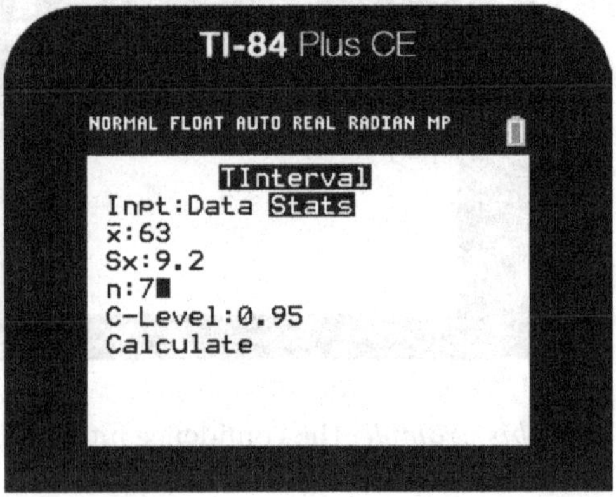

2. On the line

   `Inpt:  Data Stats`

   scroll to **`Stats`** and press enter. This will ensure

that the statistics input menu is displayed.

3. Enter the following information on the menu:

```
xbar:  63
Sx:  9.2
n:  7
C-Level:  0.95
```

   Then scroll to **`Calculate`** and press [enter].

4. The resulting confidence interval (and other information) is shown on the screen.

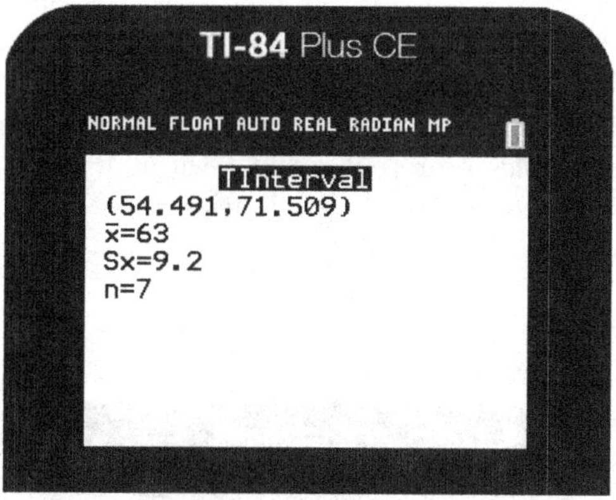

   The 95% confidence interval is [54.491, 71.509]. The center of the interval is $\overline{x} = 63$.

   The margin of error is $\pm$(71.509-54.491)/2=$\pm$8.5 points (rounded to half a point). Thus in center $\pm$ margin notation, the confidence interval is 63 $\pm$ 8.5 points.

9. 2-SampTTest – Hypothesis Test on the Difference of Two Means (from Data)

Example. It has often been stated that students who complete their basic course requirements early are more likely to graduate. Here are the graduation rates (in percent) from five colleges in California that require students to complete their mathematics requirement during their first year: 17, 16 15, 12, and 27. Here are the graduation rates from seven other colleges: 12, 14, 23, 42, 66, 72, 55. Test the hypothesis that there is a difference in graduation rates between the two populations of campuses that these samples represent against the null hypothesis that there is no difference in graduation rate.

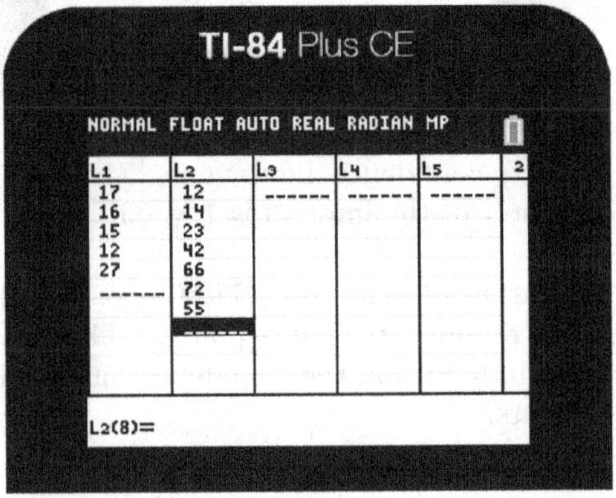

1. We will test the alternate hypothesis $H_A : \mu_1 \neq \mu_2$ against the null hypothesis $H_0 : \mu_1 = \mu_0$, using a significance level of $\alpha = 5\%$.

2. Enter the data values from the first sample into list L1, and the data values from the second sample into list L2. The screen will look something like this when you are done:

3. Select stat TESTS and scroll down to 2-SampleTTest. Press enter.

4. On the first line, that says

   Inpt: Data Stats

   scroll to **Data** and press enter before proceeding.

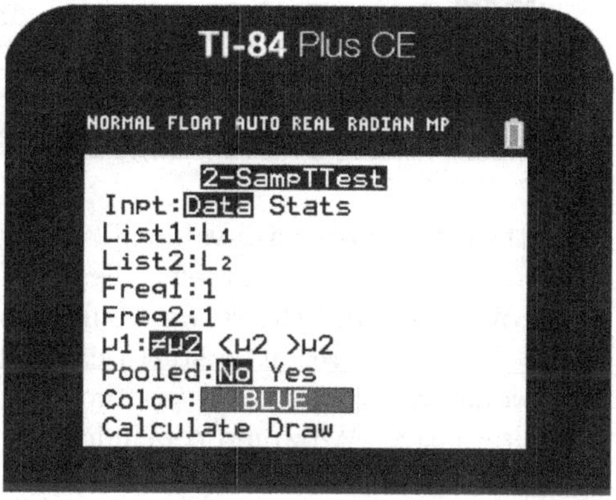

5. Enter the following information:

List1: L1
List2: L2
Freq1: 1
Freq2: 1
$\mu_1 : \boxed{\neq \mu_2} < \mu_2 > \mu_2$
Pooled: No

Then scroll to **Calculate** and press [enter].

6. The output screen will look something like this:

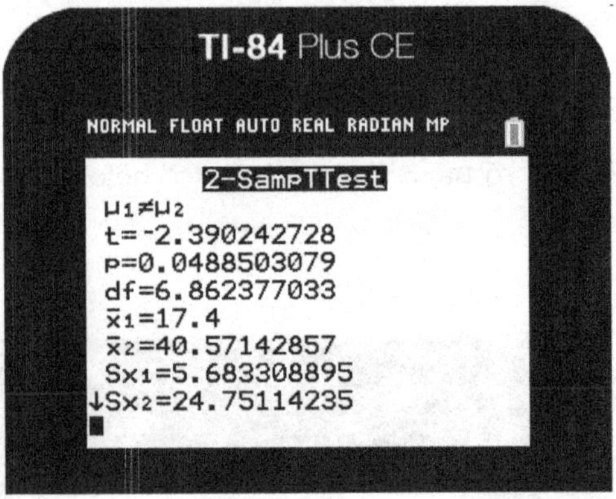

The third line shows a p-value of $p = 0.04885$, *for this example.* We have pre-specified a significance level of $\alpha = 0.05$. Hence we would observe that since $p < \alpha$ the result is statistically significant. We would reject the null hypothesis and accept the alternative. We would conclude that there is a statistically significant difference between thee represented populations at the 5% significance level.

10. 2-SampTTest – Hypothesis Test on the Difference of Two Means (from Statistics)

Example. A math teacher suspects that students who take statistics first thing in the morning will perform better than students who take the class right after lunch. He collects a random sample of student grades from last year's freshman students in the 8AM class and the 2PM class. Here are the results (the grades are measured on a 4.0 scale, where 4.0 is an A).

Class	Morning	Afternoon
Sample Size	$n_1 = 17$	$n_2 = 23$
Average Grade	3.17	2.9
Sample Standard Deviation	1.3	0.95

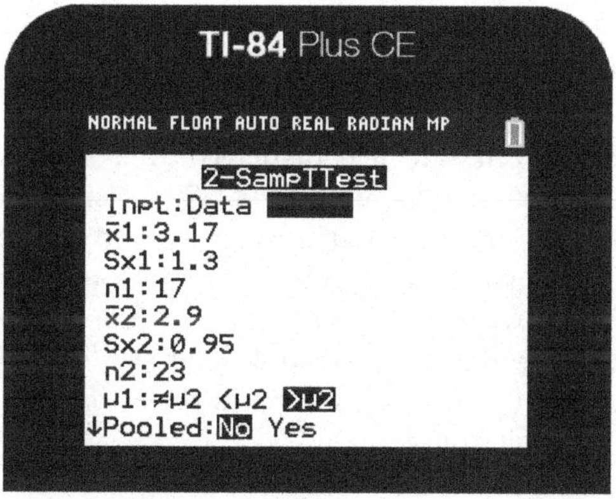

1. To test the hypothesis that morning students do better we formulate the hypotheses $H_0\colon \mu_1 = \mu_2$ and $H_A\colon \mu_1 > \mu_2$.

2. Select stat TESTS and scroll down to 2-SampleTTest. Press enter.

3. On the first line, that says

   Inpt: Data Stats

   scroll to **Stats** and press enter before proceeding. This will ensure that the statistics input menu is displayed.

4. Enter the following information on the input menu:

   $\overline{x}_1$: **3.17**
   Sx1: 1.3
   n1: 17
   $\overline{x}_2$: **2.9**
   Sx2: 0.95
   n2: 23
   μ_1 : $\neq \mu_2$ $< \mu_2$ $> \mu_2$
   Pooled: No

   The scroll **Calculate** and press enter.

5. The output screen is displayed. *For this example*, the calculated p-value is 0.2372. Using a significance level of $\alpha = 0.05$, we have $p > \alpha$ so the result is not statistically significant. We would not reject H_0 and we would not accept H_A. There is not sufficient evidence to conclude that students perform better in the morning than in the afternoon, based on this sample.

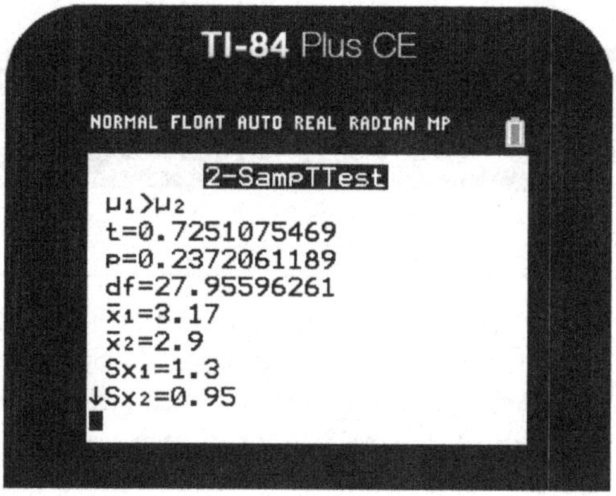
TI-84 Plus CE
NORMAL FLOAT AUTO REAL RADIAN MP
2-SampTTest
μ1>μ2
t=0.7251075469
p=0.2372061189
df=27.95596261
x̄1=3.17
x̄2=2.9
Sx1=1.3
↓Sx2=0.95

11. 1-PropZInt – Confidence Interval on a Proportion

Example. In a recent survey, 47 drivers were observed as they came to a yellow traffic light. Seventeen (17) of them stopped. Calculate a 95% confidence interval on the proportion of of drivers who stop at a yellow light.

1. Select stat TESTS and scroll down to A:1-PropZInt.... (You may have to scroll off screen to the next page.) Press enter.

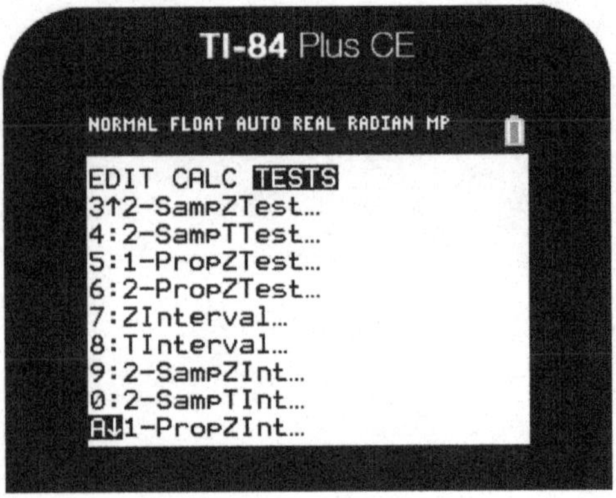

2. On the **1-PropZInt** input screen, enter the following information:

`x:`	**`17`**	number of successes
`n:`	**`47`**	sample size
`C-Level:`	**`.95`**	confidence interval

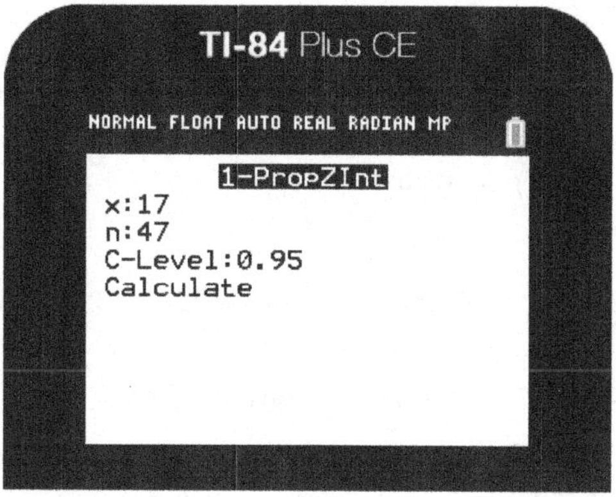

3. Scroll to **`Calculate`** and press [enter]. The output screen appears.

For this example, he confidence interval, in this case, is given in the first line as (0.2243, 0.4991).

The sample proportion is shown as $\hat{p} = .3617$.

The confidence level that was requested is shown below that (in this case, it was 95%).

To get a confidence interval in the format $p = \hat{p} \pm (ME)$, you can calculate the margin of error as either the midpoint of the interval, or the distance from $\hat{p}$ to either endpoint:

$$(ME) = (.4991 - .2243)/2 = .1374$$

or

$$(ME) = .4991 - .3617 = .1374$$

In plus/minus notation, the confidence interval is

$$p = 0.3617 \pm 0.1374$$

Since this is a proportion, we sometimes need to convert to a percent. To convert a proportion to a percent, multiply the proportions by 100. The confidence interval in percentage is

$$36.17\% \pm 13.74\%.$$

12. 2-PropZInt – Confidence Interval on the Difference Between Two Proportions

Example. Suppose researchers are comparing the effects of two different treatments for migraine headaches. Volunteer subjects in Group 1 receive *no-more-migraine* while volunteer subjects in Group 2 receive *migraine-be-gone*. Of the 73 subjects who received *no-more-migraine*, 22 reported improvement within two hours, while among the 117 subjects receiving *migraine-be-gone*, 43 reported improvement. You want to find a 95% confident interval on the difference $p_2 - p_1$, where p_2 is the (population) proportion of people who will do better with *migraine-be-gone* and p_1 is the (population) proportion who will do better with *no-more-migraine*.

1. Clearly write out the statistics and parameters:

population 1 *no-more-migraine*	population 2 *migraine-be-gone*	
$n_1 = 73$	$n_2 = 117$	number of subjects
$x_1 = 22$	$x_2 = 43$	number of successes
p_1	p_2	Population proportions
$\hat{p}_1$	$\hat{p}_2$	Sample proportions

2. Select stat TESTS and scroll down to B:2-PropZInt... .

Press enter.

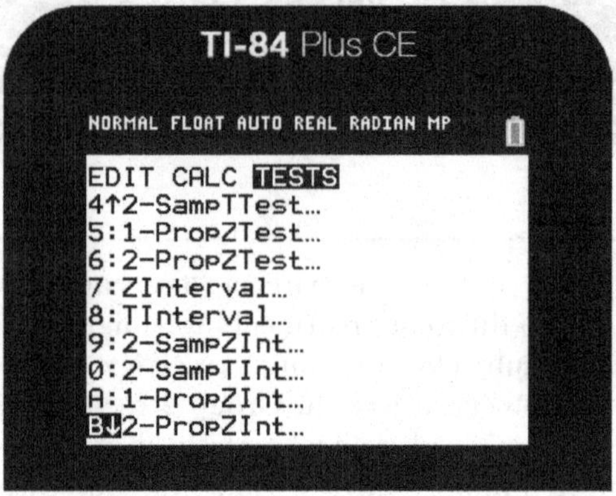

3. On the **2-PropZInt** data entry menu, enter **x1**, **n1**, **x2**, **n2** as listed above. For **C-Level**, enter **0.95** (for a 95% confidence interval).

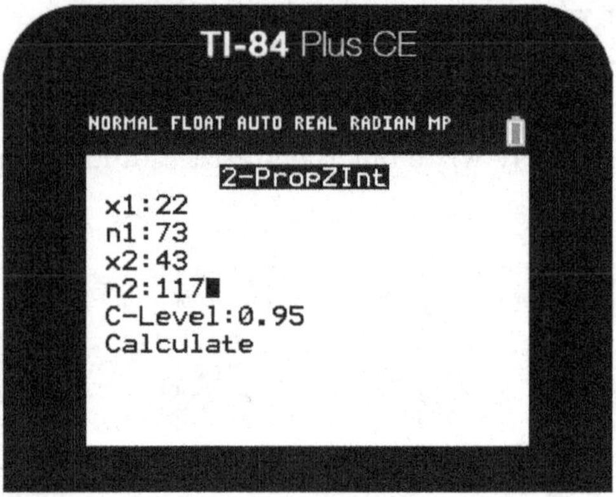

4. Scroll to **Calculate** with the arrow keys. Then press enter. The **2-PropZInt** output screen will appear.

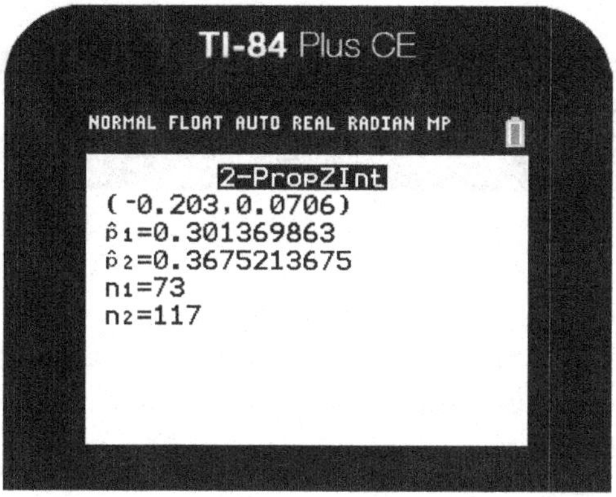

5. The confidence interval is the first item listed on the screen.

   For this example, the confidence interval is

   $$-0.203 < p_1 - p_2 < 0.0706$$

   We are also given the sample proportions, $\hat{p}_1 = 0.30136...$ and $\hat{p}_2 = 0.36752....$ The sample sizes input are also returned as n_1 and n_2.

13. 1-PropZTest – Hypothesis Test on a Proportion

Example. A recent survey states that 75% of American drivers do not come to a full stop at stop signs. You believe that the number is larger, and want to test the hypothesis:

$$H_0 : p = 0.75$$
$$H_A : p > 0.75$$

You chose a local intersection and observe that 200 of 245 drivers do not come to a full stop.

1. Select [stat] [TESTS] and scroll down to [5:1-PropZTest...]. Press [enter].

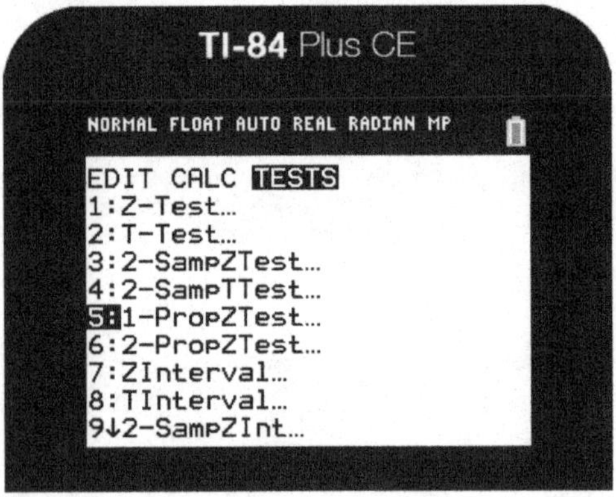

2. On the **1-PropZTest** data entry menu, enter the following information:

p0:.75	p_0 in $H_0: p = p_0$
x:200	successes
n:245	sample size
prop: $\neq p_0$ $< p_0$ $\boxed{> p_0}$	

Then scroll down to **Calculate** and press enter.

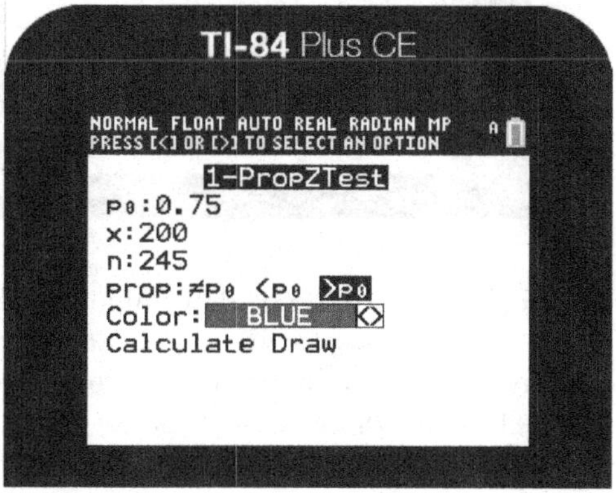

3. The **1-PropZTest** output screen will appear.

   The first line gives the alternate hypothesis H_A.

   The second line gives the z-score, $z = (\hat{p} - p_0)/SE$, where SE is the standard error.

   The third lives p, the p-value of the hypothesis test. *For this exapmle*, it is 0.008252306.

   The fourth line gives the observed sample proportion $\hat{p} = x/n$. In this case it is 0.8163265306.

The fifth line gives the sample size n. In this case it is 245.

4. The calculator does not determine whether or not your data is statistically significant. To evaluate the hypothesis, compare it with your given significance level α. *In this case*, this test has a p-value of around 0.00825. Therefore, *in this case*, one might say that the given data is significant at a p-value of 0.00825. In general, *the lower the p-value, the more significant the results.*

14. 2-PropZTest – Hypothesis Test on the Difference Between Two Proportions

Example. Suppose researchers are comparing the effects of two different treatments for migraine headaches. Volunteer subjects in Group 1 receive *no-more-migraine* while volunteer subjects in Group 2 receive *migraine-be-gone.* Of the 73 subjects who received *no-more-migraine*, 22 reported improvement within two hours, while among the 117 subjects receiving *migraine-be-gone*, 43 reported improvement. You want to test the hypothesis that *migraine-be-gone* works better than *no-more-migraine.*

1. Clearly write out the data and parameters:

population 1	population 2	
no-more-migraine	*migraine-be-gone*	
$n_1 = 73$	$n_2 = 117$	number of subjects
$x_1 = 22$	$x_2 = 43$	number of successes
p_1	p_1	population proportions

2. Define the Hypothesis:

$$H_0 : p_1 = p_2$$
$$H_A : p_1 < p_2$$

3. Select stat TESTS and scroll down to 6:2-PropZTest... .

Press enter.

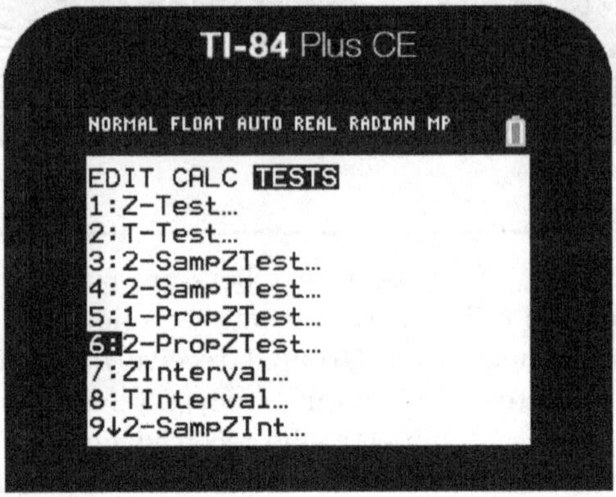

4. On the **2-PropZTest** data entry menu, enter **x1**, **n1**, **x2**, **n2** as listed above. For **p1:** scroll to **<p2**, to represent the alternate hypothesis $H_A\colon p_1 < p_2$.

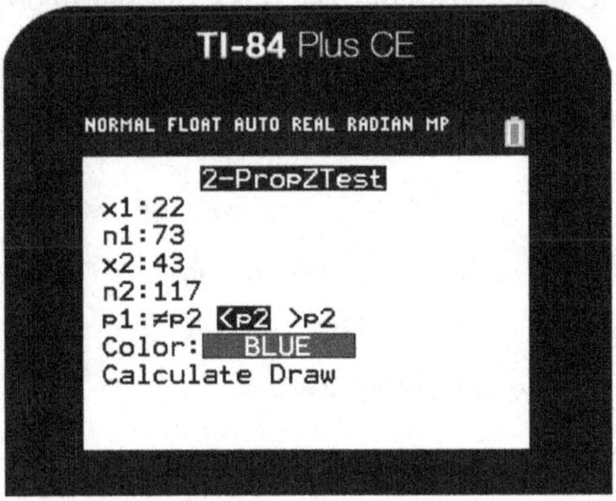

5. Press enter. The **2-PropZTest** output screen will appear.

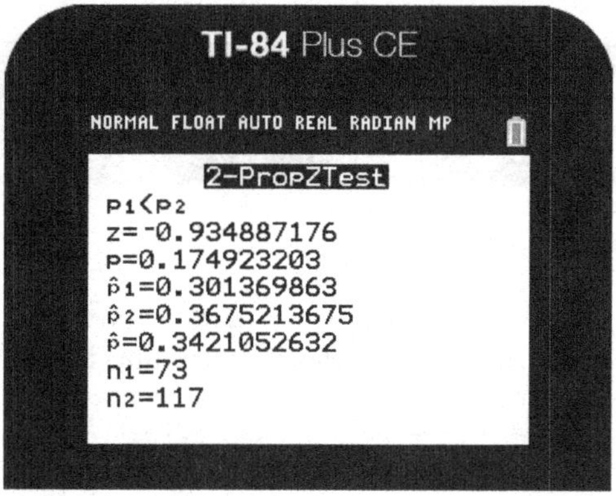

On the third line of the screen, the p-value is give. *In this case*, it is approximately $p = .1749$. The calculator does not tell you how to interpret this p-value. The determination of its significance is left to the user.

15. LinReg (ax+b) - Linear Regression

Example. Find the slope and y-intercept (linear regression of y on x) of the best fit line for the following data:

x value	1.100	1.900	2.000	2.800	3.700	4.500
y value	2.200	3.300	2.420	2.490	2.830	2.875

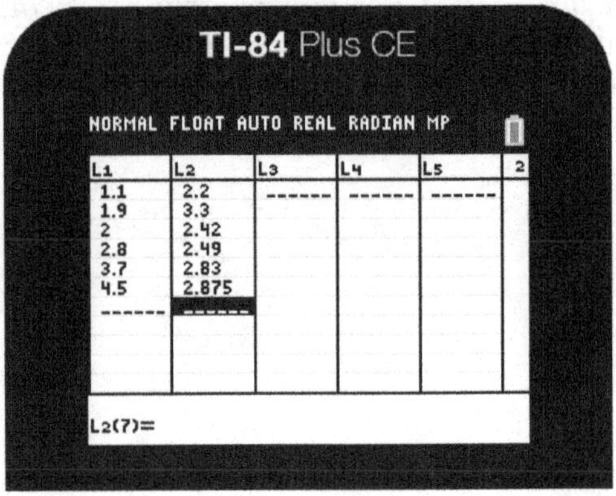

1. The equation of the regression line is given by $y = a + bx$ where a is the y-intercept and b is the the slope. We do this with the function **`LinReg(a+bx)`**. We begin by entering the data into two lists **`L1`** and **`L2`**. Put the x data into the column **`L1`** and the y data into the list **`L2`**. Do not change the order of the data - each pair of the points (each column) in the table above must be in the same row in the

table in the calculator.

2. Select stat CALC; then scroll down to **8:LinReg(a+bx)** and hit enter.

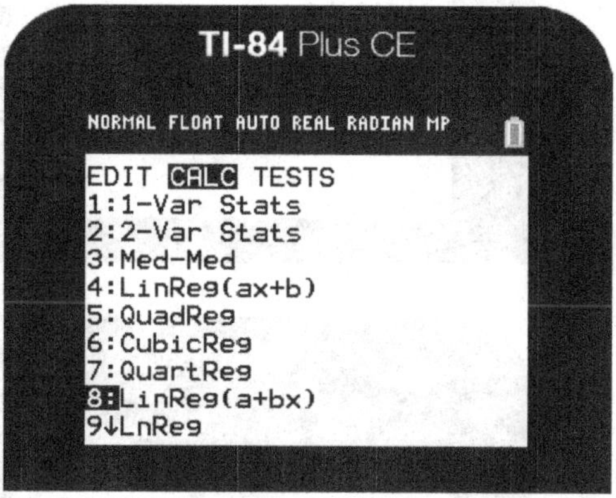

3. Enter the following information into the LinReg input menu:

XList:	**L1**	
YList:	**L2**	
FreqList:		*(Leave blank)*
Sore RegEQ:		*(Leave blank)*

Then navigate to **Calculate** and hit enter.

4. The **LinReg** results will be displayed.

For this example, the equation of the line is $y = a + bx$ where (rounded to 3 decimals) $a = 2.371$ and $b = 0.118$. The best fit line is $y = 2.371 + 0.118x$. The slope is approximately $b = 0.118$ and the y-intercept is approximately $a = 2.371$.

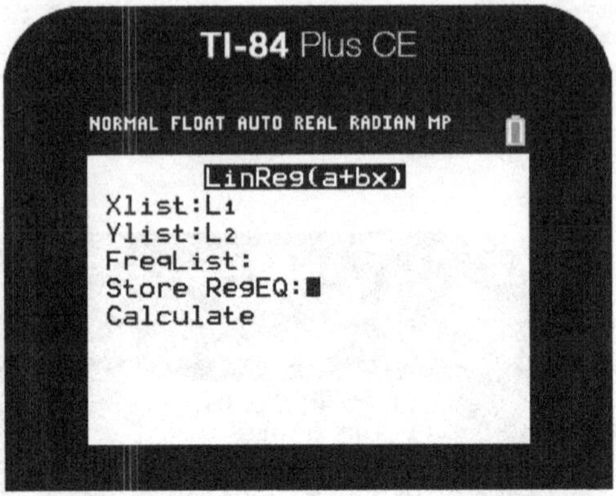
TI-84 Plus CE
NORMAL FLOAT AUTO REAL RADIAN MP
LinReg(a+bx)
Xlist:L1
Ylist:L2
FreqList:
Store RegEQ:
Calculate

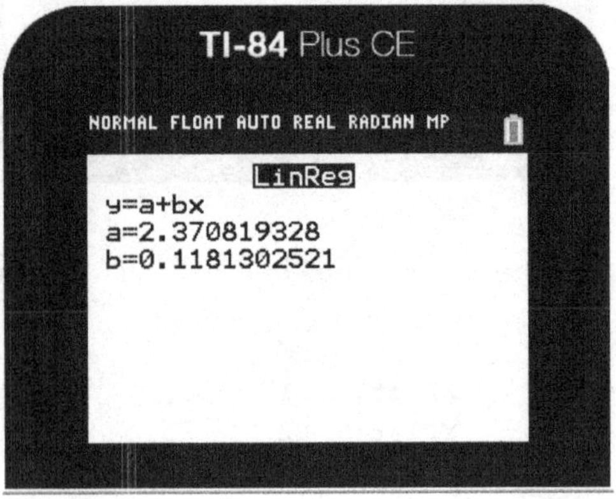
TI-84 Plus CE
NORMAL FLOAT AUTO REAL RADIAN MP
LinReg
y=a+bx
a=2.370819328
b=0.1181302521

16. Correlation

Example. Find the correlation between y and x in following data:

x value	1.100	1.900	2.000	2.800	3.700	4.500
y value	2.200	3.300	2.420	2.490	2.830	2.875

1. Select 2nd catalog. Then scroll to **DiagnosticOn**. Hit enter twice. The second time, the word **Done** will appear on the screen. You should never have to repeat this step for the life of your calculator unless (a) you reset your calculator; (b) you select **DiagnosticOff** from the catalog; (c) the battery is replaced or (d) goes completely dead.

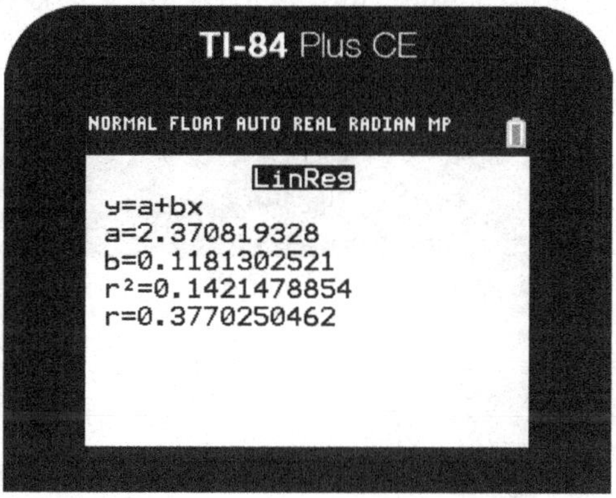

2. Repeat the linear regression procedure as described in chapter 15. The final output screen will now also display the correlation. For this data set, $r \approx 0.377$.

17. Chi Squared Test of Independence

Example. A bowl of holiday m&m's placed on a table at a party contains red, white and blue candies that are either milk chocolate or peanut filled. You are interested in determining if the color distribution depends on the type of candy, so you sample the candy. Here are your results:

	Red	White	Blue
Milk Chocolate	10	12	17
Peanut Filled	7	15	14

1. Enter the data into matrix. Hit the key sequence [2nd] [matrix]. (**matrix** is above the key [x^{-1}].)

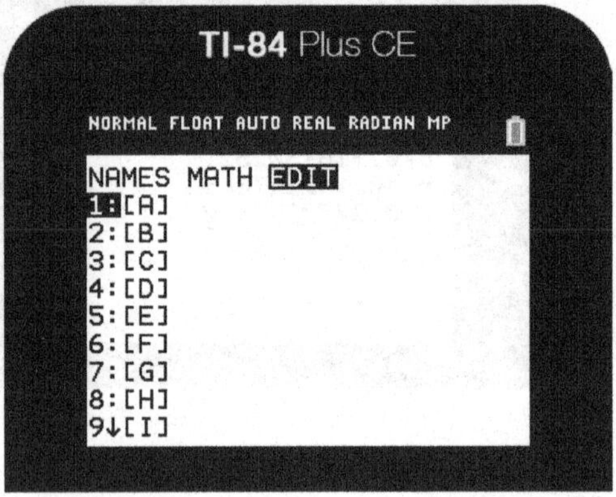

Then scroll right to **EDIT** and hit [enter] in order to edit the matrix **[A]** (If you prefer, you can select another matrix besides **[A]**.)

2. Scroll to the right and change the matrix size to **2 x 3** for 2 rows and 3 columns (you can set the size to other values for other tables). Press enter to display the blank zero-filled template.

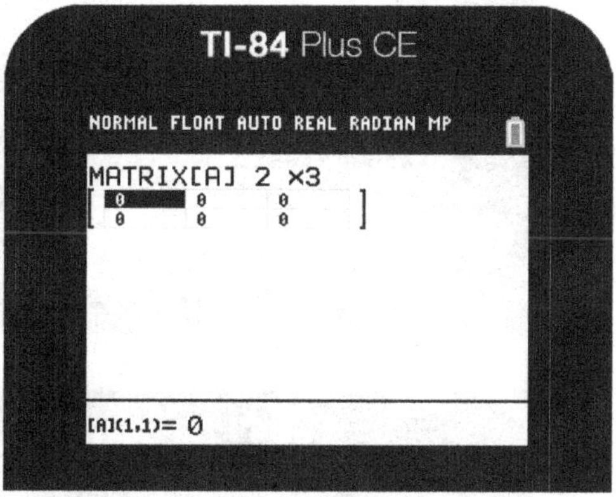

3. Copy the observed numbers from the table above into the matrix.

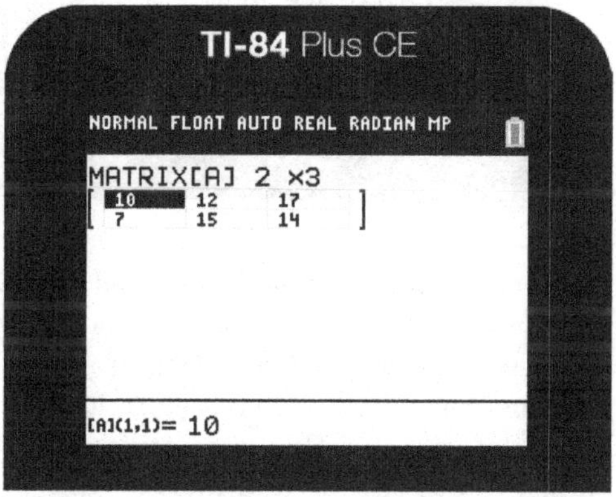

4. Click on stat, scroll right to **`TESTS`**, scroll down to **`C:`χ^2`-Test`** and hit enter.

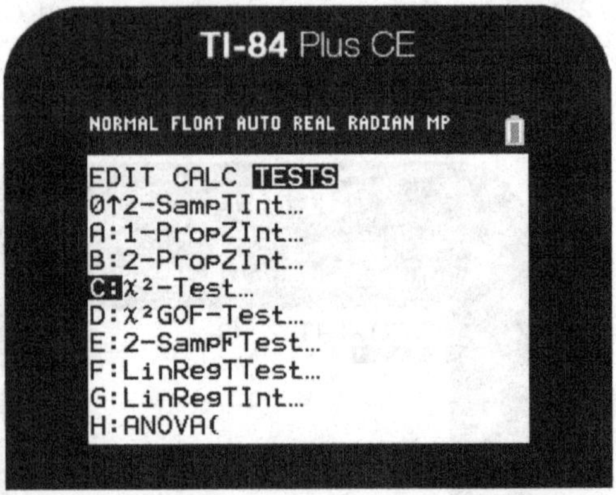

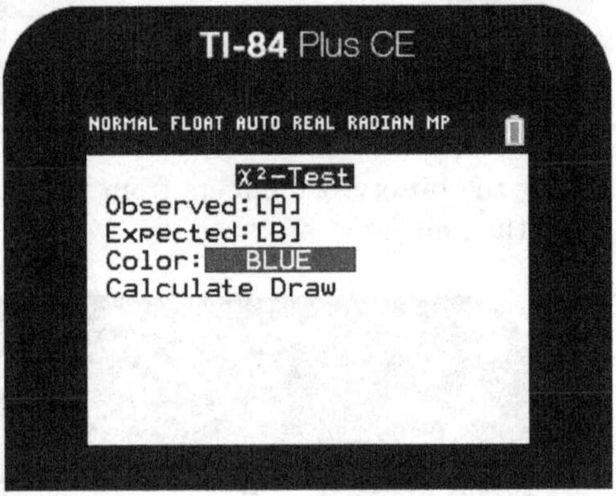

5. Select the observed and expected matrices. The observed matrix is the matrix that you just filled out (matrix **`[A]`**). The matrix of expected counts will be calculated automatically for your; the default location for this is matrix **`[B]`**. Then scroll down to **`Calculate`** and hit enter.

6. The value of χ^2 and the p-value will be displayed.

7. The predicted counts will (computed by the calculator) will be placed in matrix **[B]**. To see them, go to stats matrix then select **EDIT [B]** and hit enter.

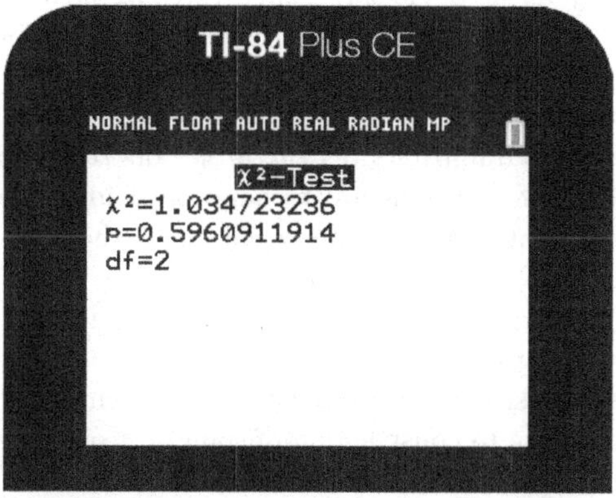

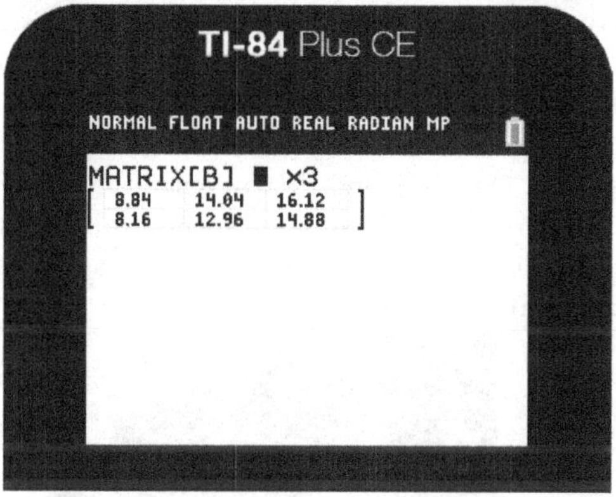

About the Authors

Bella Bluenose and Romeo Rednose are American Staffordshire Terriers who like mathematics, biscuits, running around in circles for no obvious reason, scratching themselves in odd places, and chasing the occasional squirrel. Romping around campus recently at a major California university, and hearing the pedantic rambles of stats professors through windows and besides dumpsters (as the professors searched for items to pawn for supplemental income, no doubt), they quickly realized that even without vocal cords they could explain things far more clearly than most profs. Much better, in fact, if the reek of fear and indecision coming from inside the classrooms was any indication. Lacking a universal translator (that last ones on Ebay had been purchased by a pod of dolphins due to construction of some sort of interstellar way station), and unable to find a suitable interpreter (those who know APBW, that is, American Pit Bull Woof, are way overpriced), Romeo and Bella decided to write this book. Unfortunately, computer keyboards in their arm of the galaxy are poorly matched to paws and snouts, so expect the occasional typo. And imagine the difficulty a pit bull has getting a decent editor. One bark and they're outta there! If you find an error, instead of complaining on social media, report it back to the publisher at `https://github.com/bellaromeo/faqsinstats/issues/new`.

Romeo Simon Shapiro (2007 - 2017)

Made in the USA
Monee, IL
04 November 2026

67387877R00036